The Robocentric Transhumanism Advancement
Foundation Series™

Defining Intelligence
What Artificial Intelligence Should Be

WRITTEN BY ALLEN YOUNG
NARRATED BY ALLEN YOUNG
A TRANSHUMANISTIC ASIAN-AMERICAN MAN

Robocentric.com/Books

Robocentric
Press

Copyright

Copyright © 2022 by Robocentric

All rights reserved. No part of this publication may be reproduced or transmitted in any form or by any means, electronic or mechanical, including photocopying, recording, any information storage and retrieval system, or public performance, without permission in writing from the Publisher.

"By Allen Young" on the title page of this book was rendered by the freeware Blade Runner Movie Font v3.01 created by Phil Steinschneider. The font can be freely downloaded at http://www.steinschneider.com/bladerunner/BRFont.htm.

Disclaimer

All information herein is presented for informational purposes only and offered "as is" without contract, warranty, guarantee or assurance of any kind. No responsibility or liability of any kind is assumed by the Publisher for any consequence that arises from the use of the information herein.

Table of Contents

Introduction 1
I. How Humans Use Their Intelligence 3
   1. Human Knowledge Acquisition Process 6
   2. Human Activities Process 8
   3. Human Property Production and Acquisitions 10
II. Roles of AI in Human World.....................................12
   4. AI for Human Knowledge Acquisition 13
   5. AI for Human Activities 15
   6. AI for Human Property Production and Acquisitions
....16
III. Types of Human and Artificial Intelligence 18
   7. Survival Intelligence...19
   8. Social Intelligence 21
   9. Economic Intelligence 23
   10. Sexual Intelligence 25
   11. Political Intelligence 27
   12. Competitive Intelligence......................................28
   13. Collaborative Intelligence 30
   14. Creative Intelligence 31
   15. Procreative Intelligence 33
Closing Words 34

*Defining Intelligence
What Artificial Intelligence Should Be*

Introduction

To understand the human intelligence, I look at what humans are capable of, and what humans do, in their behaviors.

Visual, aural, tactile, olfactory, and gustatory sensory information is detected by the five human senses, and enters the human brain.

The human brain processes the different types of human sensory information, creates certain emotional responses, does certain rational thinking, makes certain conclusions, makes certain decisions, and commands the human body to perform certain motor actions.

My present conclusion is that the function of the human intelligence is creating the genetically encoded human behaviors in the survival, social, economic, sexual, political, competitive, collaborative, creative, and procreative human behavior dimensions.

In my view, in developing an artificial intelligence system, trying to exactly replicate the human intelligence in machines is futile: a machine without the living biological human body will never have a form of intelligence exactly like the living human body.

If the exact replication of the human intellectual capabilities is not the end goal of AI, then what should be?

The so-called human progress and advancement is nothing more than enabling humans to have more knowledge, do more things, and have more and better things and people.

In my view, the job of an AI developer and marketer, such as my American high-tech corporation Robocentric, is to create and commercialize AI that enables human progress and advancement.

That brings the question, "How can AI play a significant, if not a paramount role, in enabling humans to have more knowledge, do more things, and have more and better things and people?"

At the highest-level design, an AI system must be designed to play vital roles in human knowledge acquisition, human activities planning and execution, and human property production and acquisitions.

This book contains some of my ideas and views on designing AI that enables the progress and advance of humankind.

I. How Humans Use Their Intelligence

Humans are never above or outside of their genetic design.

Humans are emotionally rewarded, or feel positive, good, and happy, when they do and succeed at what their genes tell them to do and be successful at.

Humans are emotionally punished, or feel negative, bad, and miserable, when they don't do and fail at what their genes tell them to do and be successful at.

In particular, humans are emotionally rewarded or punished, when they succeed or fail at the survival, social, economic, sexual, political, competitive, collaborative, creative, and procreative human endeavors.

At the physical design level, the human intelligence hardware is nothing more than the human brain with its nervous systems and the rest of the human body, that processes and applies the different types of the human sensory information, in order to emotionally reward and punish the human being according to the human genetic agenda encoded in the human individual's living human body.

One premise I use—in researching the human brain for creating the human immortality biotech and neurotech that I envision, and for advancing artificial intelligence—is that the human intelligence is nothing more than for creating the observable human behaviors and outcomes that human beings cannot help but desire and pursue, and for avoiding the human behaviors and outcomes that humans do not want.

Specifically, the human intelligence is for creating successes, and avoiding failures, in the survival, social, economic, sexual, political, competitive, collaborative, creative, and procreative human endeavors.

Humans with high intelligence succeed more in the different types of human endeavors, whereas humans with low intelligence fail more in the different types of human endeavors.

Humans, more precisely human brains, perceive the successes in the aforementioned categories of human endeavors as intelligent, and the failures in the aforementioned categories of human endeavors as unintelligent.

Humans are genetically driven to use their human intelligence—more precisely their sensory-information and stored-data processing and application capabilities in their brains and nervous systems—for creating successes in the different categories of human endeavors.

In particular, humans use their intelligence for knowledge acquisition, activities planning and execution, and property production and acquisitions—for the survival, social, economic, sexual, political, competitive, collaborative, creative, and procreative human purposes and functions.

The capability and role of AI that I envision and pursue to realize at Robocentric is augmenting the human intelligence in the fundamental human intelligence application areas, which I will talk about more in the following sections.

1. Human Knowledge Acquisition Process

The human brain is a goal-oriented sensory information processing and application system.

The human brain processes and applies the human sensory information, in order to achieve one or more human objectives in the survival, social, economic, sexual, political, competitive, collaborative, creative, and procreative human endeavors.

The human brain is capable of acquiring and applying three major types of information, which are plainly detected or observed sensory information, analyzed and deduced information, and synthesized and induced and generalized information.

With no more than these three major types of information, the human intelligence creates its successes in the survival, social, economic, sexual, political, competitive, collaborative, creative, and procreative human endeavors.

The plainly detected or observed sensory information tells the human brain the place, time, and observable characteristics of things and people.

The analyzed and deduced information tells the human brain what the different components, relationships, and causes of things and people are.

The synthesized and induced and generalized information tells the human brain how things and people work as collective bodies of systems.

The human knowledge acquisition process is going from the plainly detected or observed sensory information, to the analyzed and deduced information, and to the synthesized and induced and generalized information—for discovering new phenomena, understanding the different components and relationships and overall mechanisms of the new phenomena, and eventually applying the newly gained knowledge in creating the human successes in the survival, social, economic, sexual, political, competitive, collaborative, creative, and procreative human endeavors.

2. Human Activities Process

Humans use their intelligence to plan and execute their survival, social, economic, sexual, political, competitive, collaborative, creative, and procreative human activities.

In fact, the end purpose of the human intelligence is to plan and execute the human activities, to produce the human motor actions that result in the desired human outcomes.

In planning and executing a human activity, the human brain uses plainly detected or observed sensory information, analyzed and deduced information, synthesized and induced and generalized information, or any combination of those.

The more sophisticated and advanced the types of information used in planning and executing a human activity is, the more complex and far-reaching the outcomes and consequences of the human activity is.

The most developed and advanced human individuals, groups, organizations, societies, states, and nations use the most sophisticated and advanced types of information to plan and execute the most complex and far-reaching human activities.

The ultimate purpose of the human intelligence is to discover, create, and gather ever-increasingly more advanced human knowledge that will be used in planning and executing ever-increasingly more complex and advanced human activities!

As humans collectively advance their knowledge using their human intelligence, the human activities collectively advance as well through the applications of the accumulated human knowledge.

Ultimately, humans use their intelligence to advance their human activities.

3. Human Property Production and Acquisitions

Humans use their intelligence to produce and acquire properties—personal possessions and items, goods, services, consumables, and businesses—for their survival, social, economic, sexual, political, competitive, collaborative, creative, and procreative human endeavors.

Property production and acquisitions are the most complex human activities.

Humans devise and use tools and methods for their property production and acquisitions.

Humans form and enforce laws to regulate their property production and acquisitions.

Ever since the dawn of the Homo sapiens humankind, some 350,000 years ago, humans have been advancing their property production and acquisitions capabilities, through advancing knowledge and technology.

The end or final application of the human intelligence is advancing human property production and acquisitions.

Ultimately, as living biological beings, humans do no more than property production and acquisitions and use in their human activities.

The highest and ultimate end purpose of the human intelligence is to produce and acquire and use properties.

As more advanced human beings and their societies become, the more advanced their property production and acquisitions become.

The final and ultimate destination and application of the human intelligence is advancing human property production and acquisitions.

Humans are designed to be emotionally rewarded the most when they engage in, and advance, their property production and acquisitions.

God has created the human intelligence, ultimately for the humans to advance their property production and acquisitions.

II. Roles of AI in Human World

I strive to create the transhumanistic future of humankind that I envision, according to my plan in my book, *The Future*, particularly in America and the rest of the First World, in which artificial intelligence improves, augments, and accelerates the human intelligence in human knowledge acquisition, human activities planning and execution, and human property production and acquisition.

In my view, the role of artificial intelligence in the human world is to augment, supplement, and accelerate the human intelligence in doing what humans are designed to do using their intelligence.

At Robocentric, my transhumanistic American high-tech corporation, I strive to develop and commercialize the artificial intelligence technologies that I envision, that will assist and multiply the human intelligence to enable humans to do more humanly things.

4. AI for Human Knowledge Acquisition

For increasing the human intelligence, artificial intelligence must process visual, aural, tactile, olfactory, and gustatory sensory information in humanlike ways, and assist humans in detecting plainly observed information, and creating or deriving massive quantities of analyzed and deduced information, and synthesized and induced and generalized information.

One advantage AI can have over humans is fast and mechanical complex information processing. Although AI doesn't run on the human biological hardware, hence cannot process and create and apply information exactly like humans do, AI can excel and even surpass humans when mechanically processing and creating and applying vast amounts of complex information.

Humans are not machines: Humans can never be as good and fast and accurate as machines when performing mechanical tasks.

The human hardware, the living biological human body, is best suited for the survival, social, economic, sexual, political, competitive, collaborative, creative, and procreative human activities.

The machine hardware, made of lifeless parts, is best suited for mindless repetitions, even in information processing.

Advanced artificial intelligence must be designed in a way that augments the human knowledge acquisition capability, by performing sensory information processing in humanlike ways, mechanically detecting and applying plainly observable information, and mechanically creating and applying analyzed and deduced information, and synthesized and induced and generalized information.

In the transhumanistic human world that I strive to create at Robocentric, the role of AI is to process and apply information and create knowledge in mechanical ways with humanlike information-processing and knowledge-creation capabilities, and the role of humans is to use the information created by AI for even more humanly endeavors.

5. AI for Human Activities

To enable humans to plan and execute more human activities, especially more complex and advanced human activities, artificial intelligence must aid humans in creating and using plainly detected or observed sensory information, analyzed and deduced information, and synthesized and induced and generalized information.

Artificial intelligence must know how to detect and use plainly detected or observed sensory information for human benefit.

Artificial intelligence must be able to create and derive analyzed and deduced information, and synthesized and induced and generalized information, and use such types of information for human benefit.

Artificial intelligence must be competent in classifying and creating different types of information, and using such information in planning and executing human activities.

The ultimate role of artificial intelligence that I envision and strive to realize is performing information processing and creation and application in humanlike, albeit mechanical ways, to aid humans in planning and executing human activities.

6. AI for Human Property Production and Acquisitions

I strive to develop and integrate artificial intelligence into every facet and fabric of the human property production and acquisition activities, to increase and accelerate human property production and acquisitions.

In my vision of the transhumanistic future of humankind that I strive to realize, particularly in America and the rest of the First World, the central role of artificial intelligence is to enable humans to produce, acquire, and consume more properties, not just on Earth, but more importantly in outer space.

Artificial intelligence must be used in every step of human property production and acquisition planning and execution.

Artificial intelligence must have enough capability to be useful in every part of human property production and acquisition planning and execution, for enabling humans to produce, acquire, and consume better and more properties.

Artificial intelligence must aid humans to be far more prolific property producer and consumer, on Earth and in outer space.

Artificial intelligence must enable humans to create and consume more and better for even more advanced survival, social, economic, sexual, political, competitive, collaborative, creative, and procreative human activities.

III. Types of Human and Artificial Intelligence

Humans are surviving, social, economic, sexual, political, competitive, collaborative, creative, and procreative creatures.

To be human is to be surviving, social, economic, sexual, political, competitive, collaborative, creative, and procreative.

The human intelligence makes humans surviving, social, economic, sexual, political, competitive, collaborative, creative, and procreative.

In this part, I'll take a closer look at how the human intelligence is used in the different dimensions of the human behavior, and how artificial intelligence should be used to augment each human behavior dimension.

7. Survival Intelligence

To survive, humans must eat, clothe themselves, keep themselves in a shelter, and keep themselves out of bodily physical harm. The human intelligence enables humans to do all these things that are required for the human survival.

Humans are natural-born biological survival machines.

While humans can survive just fine on Earth without artificial intelligence, the humans cannot so readily survive in outer space without the help of artificial intelligence.

In artificial nuclear-fusion powered interplanetary spaceships and mass-scale outer space human habitats, artificial intelligence must constantly monitor the surrounding environments using sensors to ensure the human survival in outer space by constantly providing the necessary conditions for outer-space human survival.

Outer space is an extremely dangerous and hostile place to humans, with so many things that can go wrong and jeopardize human survival in outer space, especially in a long-term human survival in outer space that will last days, weeks, months, years, decades, centuries, millennia, and beyond.

Most people who leave Earth will never come back to Earth.

In the future, most humans will not be born on Earth; most humans will be born in outer space.

In order to ensure the survival of humans in outer space, artificial intelligence is needed.

Artificial intelligence will be used in ensuring the survival of humankind in outer space in massive numbers!

8. Social Intelligence

Humans are powerful, because humans are social.

Humans are social, not solitary, creatures that form relationships with, work with, and rely on other human beings to create the sum of human beings that is far greater than the individual human beings.

The human intelligence enables humans to band and work with other human beings.

The role of artificial intelligence in the social human settings is to enable humans to band and work with other human beings in even more advanced and complex ways, on Earth and in outer space.

Artificial intelligence must enable humans to form even more advanced human relationships and work together for greater mutual human benefit.

Artificial intelligence must be a technology that enables vastly more complex and advanced social human activities.

Artificial intelligence must enable humans to plan, execute, monitor, and improve immensely sophisticated and advanced social human activities on Earth and in outer space.

Artificial intelligence is for creating even more complex social human activities on Earth and in outer space!

9. Economic Intelligence

Humans are powerful, because humans produce and consume.

Humans harvest and transform resources to create goods and services that are beneficial to them.

The human intelligence enables humans to plan and execute their production and consumption activities.

The role of artificial intelligence in the economic human activities is to enable humans to produce and consume more and better on Earth and in outer space.

Artificial intelligence must be used in every stage and step of economic production and consumption human activities.

Artificial intelligence must enable humans to vastly increase and further diversify what humans produce and consume, particularly in America and elsewhere in the First World.

I want artificial intelligence to exponentially increase the human economic activities, particularly in America and elsewhere in the First World.

As I have publicly stated numerous times, one key goal I pursue in transhumanizing America is doubling the

American GDP by making AI and robots ubiquitous in America.

Robocentric's development and commercialization of its artificial nuclear-fusion powered mass scale outer space humanity expansion technologies will enable humans to harvest, process, and consume the astronomically vast amounts of material resources in outer space.

Artificial intelligence is needed for enabling and managing harvesting and processing the vast outer space resources.

The economic artificial intelligence that Robocentric works on developing and commercializing will have the capability to increase the economic production and consumption human activities and capabilities.

Artificial intelligence is for increasing the economic production and consumption human activities and capabilities on Earth and in outer space!

10. Sexual Intelligence

Humans are sexual creatures by nature, by genetic design.

To be human is to be sexual.

The human intelligence enables humans to find one or more sexual partners, and form and maintain one or more sexual human relationships.

Some humans are more successful at sexual human activities than other humans.

The role of artificial intelligence in sexual human activities is helping humans to be more successful in their sexual activities.

Artificial intelligence must be able to help humans to be more sexually desirable.

Artificial intelligence must be able to help humans maintain their existing and new sexual relationships.

Artificial intelligence must enable humans to be more sexual.

Artificial intelligence must enable humans to be more successful in sexual mate finding, and sexual relationship maintenance.

Artificial intelligence is for enabling humans to be more successful sexual creatures on Earth and in outer space!

11. Political Intelligence

Humans band with other humans for their common, shared benefit through setting and enforcing the rules for governing themselves: Humans are political creatures.

The human intelligence enables humans to create and enforce governing rules that benefit themselves.

The role of artificial intelligence is to help humans to govern themselves better.

Artificial intelligence is for enabling humans to maximize their self-beneficial self-governing.

Artificial intelligence is for aiding humans in planning and executing human political activities.

Artificial intelligence is for helping humans in planning and executing human political campaigns and political public communications, and forming and managing political organizations.

Artificial intelligence is for enabling humans to be more effective and efficient in governing themselves for their benefit.

Artificial intelligence is for enabling humans to be more powerful political creatures on Earth and in outer space!

12. Competitive Intelligence

The human desires to be better than other human individuals, couples, families, groups, organizations, societies, and nations compel humans to evolve and advance.

The history of humankind has been a history of human progress and advancement via human competition.

The perpetual human desire to be better than others puts the human intelligence to work to create more of better things, and to be more successful and dominant than others.

Artificial intelligence must cater to the competitive human desires.

Artificial intelligence must enable humans to create and have more of better things.

As technology products, artificial intelligence must give its human users a competitive edge in complex information processing, creativity, human activities management, and organization management.

Artificial intelligence already is a human competition tool: it will only become more so.

In the world I envision and strive to create, the humans that use advanced artificial intelligence will have a tremendous competitive edge over others who don't use artificial intelligence.

Artificial intelligence is a technological tool: The ultimate function of a technological tool is giving its user a competitive advantage in accomplishing his or her tasks.

At Robocentric, I strive to develop and commercialize the artificial intelligence technologies that I envision that will give their users competitive advantages over others who don't use the Robocentric artificial intelligence technology products.

13. Collaborative Intelligence

The human intelligence compels humans to seek, band with, and work with others for mutual benefit, because collaboration is a smart thing to do for humans.

How can artificial intelligence enable humans to collaborate with other humans better?

Humans work together; artificial intelligence must enable humans to work together better and faster.

The future artificial intelligence that I strive to develop and commercialize at Robocentric must be able to compute and predict the optimal methods of human collaboration, by computationally modeling and simulating human brains.

That is, artificial intelligence must serve as a human collaboration optimization technology, on Earth and in outer space.

Humans collaborating with each other more and better using artificial intelligence is one aspect of the transhumanistic future that I strive to create.

14. Creative Intelligence

As I stated earlier, the ultimate end and purpose of the human intelligence is to create.

In order for artificial intelligence to enable humans to be more creative, it behooves to look into the human creativity process.

Humans create by conceiving a need or want, finding and/or developing one or more tools and methods for creation, then developing and executing a process for creation by trial and error.

Artificial intelligence can play a vital role in the human creation process, by automating certain aspects of creation tools and methods finding and development, and creation processes development and execution.

At Robocentric, I pursue maximum automation by artificial intelligence.

I strive to create the future human reality in which machines will do all the uncreative, mechanical work, so that humans can focus exclusively on doing creative work and being more human.

Total elimination of the human need to do repetitive, mindless, and mechanical work that machines can and should do is one dimension of the transhumanistic future that I strive to build.

15. Procreative Intelligence

Humans spend enormous amount of time and energy on procreating and raising their offsprings.

Artificial intelligence should help humans in the procreative human endeavors, by automating all the human child rearing tasks that can be automated, such as cooking at home and washing the kids' clothes.

In other words, artificial intelligence should do all the mechanical child-rearing tasks done by humans, so that humans can focus exclusively on interacting with their children, and do none of the repetitive mechanical chores that have been done by humans so far in child rearing.

Closing Words

The human reality happens according to the human genetic design.

Humans are genetically designed for the survival, social, economic, sexual, political, competitive, collaborative, creative, and procreative human activities that humans have performed so far, and will continue to perform until humans are genetically engineered and neurotechnologically redesigned.

In the not-too-distant future, I hope to realize and commercialize the artificial intelligence technologies that enable humans to do more of the human activities that they are genetically designed to do.

To have fun as human beings is to do what humans are genetically designed to do: I will keep on pursuing developing and commercializing the artificial intelligence technologies that I envision, which if and when realized, will enable humans to have a lot more fun through using intelligent machines that can do a lot of boring, repetitive, dull, and mindless tasks for humans, so that humans can focus on doing creative and managerial work.

About the Author

Allen Young is a transhumanistic Asian-American man who publicly promotes and advances AI, robotics, human immortality biotech, and artificial nuclear-fusion powered mass scale humanity expansion tech.

Allen Young is currently working on raising capital for his high-tech startup, Robocentric, and completing and commercializing his visual AI and robotics software technologies.

More information about Robocentric is at Robocentric.com/About.

You can invest in Robocentric during its early startup years at Robocentric.com/Investors.

A list of all of Allen Young's books on transhumanism with purchase links is available at Robocentric.com/Books.

(The end)

This has been 'Defining Intelligence: What Artificial Intelligence Should Be', Written by Allen Young, Narrated by Allen Young, Copyright 2022 by Robocentric, Production Copyright 2022 by Robocentric. No part of this publication may be reproduced or transmitted in any form or by any means, without permission in writing from the Publisher.

www.ingramcontent.com/pod-product-compliance
Lightning Source LLC
Chambersburg PA
CBHW050318220526
45465CB00005B/2038